COMMON MAN
★★ UNCOMMON LIFE ★★

FIELD MANUAL

COMPANION GUIDE TO
COMMON MAN | UNCOMMON LIFE
VIDEO SERIES

COPYRIGHT © 2015 | LIGHT THE DARK, INC. • JACKSONVILLE, FL
Design by Alex Watson Creative, LLC. • Written by Stephen Freeman
Contributors: Jeff Bramstedt, Stephen Freeman, Henry Mullis, Ian Otto

Scripture quotations marked (NIV) are taken from the Holy Bible, New International Version®, NIV®.
Copyright © 1973, 1978, 1984, 2011 by Biblica, Inc.™ Used by permission of Zondervan.
All rights reserved worldwide. www.zondervan.com The "NIV" and "New International Version"
are trademarks registered in the United States
Patent and Trademark Office by Biblica, Inc.™

Scripture quotations marked (ESV) are from The Holy Bible, English Standard Version® (ESV®),
copyright © 2001 by Crossway, a publishing ministry of Good News Publishers. Used by permission.
All rights reserved.

★ TABLE OF CONTENTS ★

NAVY SEAL ETHOS

INTRODUCTION

EPISODES ★

1 I AM THAT MAN (1)

2 MY TRAINING IS NEVER COMPLETE (13)

3 I AM NEVER OUT OF THE FIGHT (25)

4 LEAD & BE LED (37)

5 MY CHARACTER & HONOR ARE STEADFAST (49)

6 I HUMBLY SERVE (61)

7 I WILL GET BACK UP, EVERY TIME (73)

NAVY SEAL ETHOS

In times of war or uncertainty there is a special breed of warrior ready to answer our Nation's call. A common man with uncommon desire to succeed. Forged by adversity, he stands alongside America's finest special operations forces to serve his country, the American people, and protect their way of life. *I am that man.*

My Trident is a symbol of honor and heritage. Bestowed upon me by the heroes that have gone before, it embodies the trust of those I have sworn to protect. By wearing the Trident I accept the responsibility of my chosen profession and way of life. It is a privilege that I must earn every day. My loyalty to Country and Team is beyond reproach. *I humbly serve* as a guardian to my fellow Americans always ready to defend those who are unable to defend themselves. I do not advertise the nature of my work, nor seek recognition for my actions. I voluntarily accept the inherent hazards of my profession, placing the welfare and security of others before my own. I serve with honor on and off the battlefield. The ability to control my emotions and my actions, regardless of circumstance, sets me apart from other men. Uncompromising integrity is my standard. **My character and honor are steadfast.** My word is my bond.

We expect to lead and be led. In the absence of orders I will take charge, lead my teammates and accomplish the mission. I lead by example in all situations. I will never quit. I persevere and thrive on adversity. My Nation expects me to be physically harder and mentally stronger than my enemies. ***If knocked down, I will get back up, every time.*** I will draw on every remaining ounce of strength to protect my teammates and to accomplish our mission. ***I am never out of the fight.***

We demand discipline. We expect innovation. The lives of my teammates and the success of our mission depend on me - my technical skill, tactical proficiency, and attention to detail.
My training is never complete. We train for war and fight to win. I stand ready to bring the full spectrum of combat power to bear in order to achieve my mission and the goals established by my country. The execution of my duties will be swift and violent when required yet guided by the very principles that I serve to defend. Brave men have fought and died building the proud tradition and feared reputation that I am bound to uphold. In the worst of conditions, the legacy of my teammates steadies my resolve and silently guides my every deed. I will not fail.

WELCOME TO
COMMON MAN | UNCOMMON LIFE

The Field Manual is a companion book that goes along with the videos you'll watch with your group each week. This guide will help you engage with the content on a personal level and reflect on what you've seen and heard each week.

The content in your Field Manual pulls some thoughts from the Navy SEAL ethos but the roadmap for the series is the Word of God. Over the course of the series, we'll discuss timeless Biblical principles that Jesus says leads to an adventurous, uncommon life. Ultimately, we hope that this manual helps you move closer to God and encourages you to join the mission that He has for you.

HOW THE SERIES IS SET UP

There are five lessons for each week in this series and they are designed to be completed throughout the week. There are reflection questions and spaces to write, so we hope that you honestly and authentically engage with the content. **Each day should take no longer than 10 minutes to complete.**

WHAT YOU WILL NEED

Pen or Pencil – Although the Field Manual is not heavy on writing responses, there are some sections to write in. Make sure you have something to write with.

Bible – For the most part, the verses for each day are printed in the content, but it's a good idea to have one nearby as needed.

Time – Our goal is to streamline the content so that each day is impactful inside a limited window of time. Our intention is that each day would take less than 10 minutes to complete.

Quiet Space – God often speaks when we are making the time to listen, so try to put other distractions on hold.

It is our honest prayer that God uses this series to impact your life and other men's lives all over the world. We hope this Field Manual is an asset to your life and is something you can refer back to long after this series is over. Let's do this together and find out what God says it takes for a Common Man to lead an Uncommon Life.

I AM THAT MAN

DAY ONE
EARNED VS. GIVEN

Becoming a US Navy SEAL is something that is earned and worked for – not something that is just given away freely. It's important to make the distinction between things that we earn and things we are given. In Romans 6:23 (NIV), it says:

> *"For the wages of sin is death, but the free gift of God is eternal life in Christ Jesus our Lord."*

A "wage" is something you work for. It is the fair compensation for your labor and it is earned, not gifted. You are cheated if you work for something and are then denied it.

In the verse above, it tells us that our full-time job was sinning, and the wage for that work is death. Because God is perfectly Just, what we rightly earned had to be paid out. If you are paid what you earned, you should be put to death.

JUSTICE AND MERCY
God is not only Just, He is also Love. So He devised a plan from the very beginning that would fulfill His nature of Justice as well as His nature of Love. He would stand in the gap between death and us, so that the wage we should have received would still be paid, but that we wouldn't have to be the ones to pay it.

THE FREE GIFT
The full wages for your sin was given to Jesus on the cross, and He took them on your behalf. And what is the price He charged us to do that? Nothing. The gift of eternal life is yours – it can't be earned or paid for or bought or stolen. Jesus paid the entire cost, and He gives eternal life away for free – extravagantly – to anyone who accepts it.

MIXING UP THE TWO

Where a lot of people get into trouble, is when they forget what is earned and what is given freely. Believing that God gave you eternal life because of your righteousness is worse than a lie. We have not earned a place in the Kingdom of God, and we never could. Eternal life is a gift that cannot be bought, and death is a wage that could never have been escaped.

People who try to be perfect and sinless on their own can never earn their way out of the wages of death, but those who admit they can't do it, humble themselves, and simply accept the free gift of Jesus will enter the Kingdom of Heaven.

The Challenge this week is simple. Have you accepted the free gift of salvation? If not, write down what is holding you back?

_____

_____

_____

_____

_____

_____

WALK OUT ALIVE

The first step is to just accept the gift He's given, but it doesn't stop there. If your legs are no longer paralyzed, then you're free to run and jump and climb like you couldn't before. Being free from the crushing debt of sin is not the end of a story; it's the healing that starts the very first chapter.

DAY TWO
THE PAST DOES NOT DISQUALIFY YOU

Jeff mentioned in the video this week that many people look at what Jesus did on the cross and have a hard time believing that this level of forgiveness extends to them. If you are struggling with a past of drugs or alcohol abuse, sexual sin, or murder, it might be hard for you to believe that a free gift from God could really wipe out the sin in your past. After all, did God really have your past in mind when He chose to pay for the sin of the world?

THE REASON FOR REDEMPTION
Although the actual application of the idea may be difficult for you, the truth is that this grace absolutely extends to you. Psalm 139:4 (ESV) says:

> *"Even before a word is on my tongue, behold, O Lord, you know it all together."*

Even if your past is so terrible that you can't even talk about it – God knew then, He knows now, and He still chose to love and forgive you.

PUTTING IT IN PERSPECTIVE
One outstanding example in the Bible is in the story of King David. Throughout scripture, David is known as a man after God's own heart and is revered as the greatest and most respected king in the history of Israel.

But David was not without fault. In fact, he is guilty of sins that many people would consider beyond forgiveness. Not only did he commit adultery with Bathsheba, but to cover up the pregnancy that resulted, he had her righteous husband killed at the frontlines of battle. This certainly doesn't sound like the biography of a man of God.

But it doesn't stop with David. Moses was guilty of murder. Paul (formerly Saul) persecuted, imprisoned, and condoned the murder of early Christians. Rahab was a prostitute. Peter openly denied Christ and abandoned Him, swearing that he never even heard His name. Jacob swindled his brother's birthright from him. Jonah knew that God was going to forgive Nineveh, and because he hated them so much, ran in the opposite direction from where God told him to go. Adam and Eve disobeyed the one rule they had and separated mankind from God. The Bible is packed full of men and women, just like you and those you know, who failed on massive scales, but were still greatly used by God.

THE GREATNESS OF GOD'S MERCY
When Jesus died on the cross, it was to build a bridge between us and God. Jesus loved you so much and decided that you were worth His very life.

Believing that your past or sin is too great for God does two things at the same time. It gives you far too much credit and it gives God far too little. It grossly underestimates God's capacity for love and it dramatically exaggerates your ability to create something that is too much for God. After thinking through the greatness of God's mercy, can you name one thing in your past that you think God can't redeem?

_____

_____

_____

_____

_____

At some point today, take 3 minutes just to thank God for wiping away the sin in your past. Be real with God and don't hold back.

DAY THREE
WHAT DOES "THAT MAN" LOOK LIKE?

The Navy SEAL ethos has a challenge that says, "I am that man." For SEALs, this simply means that they are willing to accept the responsibility of upholding all the values instilled in them. As a Christian, Jesus was "That Man" for us. He took the responsibility for the payment we couldn't make and did it Himself. If you are going to be "That Man" in your own life, you have the responsibility of following Jesus and every day striving to become more like Him. It's important that we have a clear goal of character qualities of the man we are striving to be.

A WAY OF LIFE
Every person who has accepted God's gift of eternal life has been transformed into a new creation *(2 Cor. 5:17)*, and continues to be changed into God's likeness *(Phil. 1:6)*. This is a lifelong calling of all believers– to recognize the change that has **already happened** and to embrace the change that **continues to happen.**

FOCUSING YOUR ATTENTION
Below is a chart of some of the character qualities of "That Man." Go through the list below and circle character qualities you know you need to work on in your daily life, and underline the qualities you believe are already true of you.

Try substituting "That Man" with your first name, and read through this list to hear if it is true, and think of ways you can become more like the man God has called you to be.

THAT MAN:

Is ready to answer the call	Has Self Control	Fights for the Cause of the Persecuted	Never Boasts or Brags	Stands alongside Others
Recognizes Those Who Have Gone Before	Is a Protector of Those in Need	Is Responsible	Is Loyal	Is Humble
Forgives His Enemies	Does Not Seek Attention	Is Obedient and Respectful of Authority	Puts Other's Needs Before His Own	Expects to Serve, Not to Be Served
Loves Justice and Seeks Mercy	Is the Same in Private as He is in Public	Is of Good Character	Is True to His Word	Spends Time Alone With God
Is Patient and Kind	Is Focused on the Kingdom of God	Leads by Example in All Situations	Prays Without Ceasing	Does not Gossip or Slander Others
Is Resilient Even after Defeat	Is Generous With What He Has	Is Disciplined	Is Intentional	Is Faithful in Small Things
Uses His Skills to Further the Mission	Is Always Teachable	Controls His Pride	Is a Man of Action	Never Compromises His Values

DAY FOUR
HOW DO I BECOME LIKE "THAT MAN"?

The list of characteristics from yesterday can be helpful, but if your goals don't translate into action, they're meaningless. Let's look at some practical tools to help you do something today and move closer to being more like Jesus.

MACRO-VISION/MICRO-VISION
Accomplishing a goal requires two points of view. The first is "Macro-Vision," which is the big picture goal you are trying to accomplish. The second is "Micro-Vision," and this is made up of the small activities that are building toward the larger goal. They work together like a mosaic picture – the small pieces in the right places make the larger image clear.

SETTING UP THE BIG PICTURE
First, map out your large goals. These might be things like, Stop Looking at Pornography, Be a Better Father/Husband, or Have a Consistent Devotional Time. You could also work from an item on the list from yesterday. Write down three big picture goals:

1 _____
2 _____
3 _____

FAITHFUL IN SMALL THINGS
Now that you have your goals laid out, start mapping out some small steps you could take – today – to move toward accomplishing them. It's okay that these steps are very small. In Luke 16:10 (NIV), Jesus says:

"Whoever can be trusted with very little can also be trusted with much, and whoever is dishonest with very little will also be dishonest with much."

The enemy will tell you that the little things don't matter, but lives hinge on small habits, repeated over time, and formed into muscle memory.

Remember, you are always becoming someone – make certain that it's the man you want to be.

Look at the three larger goals you wrote down and name one thing for each that you can do today that would be a step toward reaching it. These could be things like, Keep My Computer in a Public Place, Count to 10 Before Responding in Anger, or Go to Bed Earlier. Then, be faithful in each small step, and see what happens.

1 _____
2 _____
3 _____

A DAILY BLUEPRINT

If it helps, try thinking of today as the blueprint that you are writing for every day for the rest of your life. You are tracing the outline of the life you want to live, and the more often you trace it, the firmer those lines become.

So if you don't want to start everyday by rolling over and checking your newsfeed on your phone, don't do it today. Character and discipline is built over time, but it has to start by doing the right thing just once.

At some point today, accomplish at least one of the small goals you wrote down. Redeem the time for something valuable while you still have it to redeem.

DAY FIVE
WHAT IS HOLDING YOU BACK?

In Jeff's challenge at the end of the video this week, he asked you to name anything that could be holding you back from following Jesus. Whether that means accepting His gift for the first time or deciding to be intentional and active in your faith, it's critical to put a name to anything that is getting in your way.

DEVELOPING YOUR STRATEGY
To accomplish anything, you need at least four things. Missing even one of these things, could seriously impair your effectiveness:

1. **A Goal:** Knowing exactly what you are trying to accomplish.
2. **Motivation:** The deep and active desire to accomplish your goal.
3. **Foresight:** Seeing what could get in your way.
4. **Strategy:** Having a plan for overcoming those roadblocks.

You have already set out your Goals, and they should be goals you are Motivated to accomplish. Your Foresight and Strategy may be apparent right away, but might need time, prayer, or the help of someone else.

THE WISDOM OF ASKING
Hindsight may be 20/20, but wisdom says to use the collective experience of those of who have gone before you as well. Proverbs 15:22 (ESV) says:

"Without counsel plans fail, but with many advisers they succeed."

You are not the first person to face these roadblocks, so find other men who have successfully reached these goals, and find out what roadblocks you should be on the lookout for.

GOAL:

MOTIVATION:

POTENTIAL ROADBLOCK:

STRATEGY:

GOAL:

MOTIVATION:

POTENTIAL ROADBLOCK:

STRATEGY:

WHAT IS AT STAKE?

Get into the habit of naming what your goals are and what is at stake if you don't meet them. God has changed the world countless times by using common men who made themselves available to Him. ***Remember that your life and the man you are becoming matters – to you, to the people counting on you, and to the Kingdom of God.***

MY TRAINING IS
NEVER COMPLETE

DAY ONE
BACK TO THE BASICS

This week, Jeff talked about the importance of training in your spiritual life, and used SEAL training as an example. Training is the disciplined repetition of very basic moves, which builds a foundation for everything else you do. Whether it's physical or spiritual, training isn't something you do once and never return to. If a SEAL made it through BUD/S training and never did a pushup again, all that preparation would be quickly negated.

BODYBUILDING VS. TRAINING
When training your spiritual muscles, remember that you are training for combat, not a bodybuilding competition. It's not so you can know more or be a better Christian than someone else. You are training so that when you are face-to-face with the enemy, you will have the skills deeply ingrained, so there is nothing but swift action.

Are there any people you know who are exceptionally disciplined in their spiritual training? Write down their names and what habits they do well. This week contact at least one of them and schedule time to talk with them about that habit.

SPIRITUAL DISCIPLINES

Below is a chart of some basic spiritual disciplines. Remember, you're simply trying to build spiritual fundamentals into your life. Circle the ones you need to work on and underline the ones you think you've successfully integrated into your natural habits already.

Praying	Fasting	Memorizing Scripture	Sharing Your Faith
Studying the Word	Waiting on the Lord	Dying to Your Old Nature	Listening to God's Voice
Resting	Giving	Serving	Holding Your Tongue
Worshipping	Fellowshipping	Teaching Others	Being Held Accountable
Confessing Sin	Praising God	Sharing Your Testimony	Putting Others First

Look at the disciplines you circled. What steps can you take today that will help you put them into practice? **Remember, since you are training for something, your training matters.**

DAY TWO
DAILY DISCIPLINES

When SEALs train, they are developing highly refined muscle memory. And you are no different. You are always training your automatic responses – for good or for bad. The question is, are you developing the right kind of muscle memory?

THE MYTH OF THE TRICK
There is something inside us that wants to know the secret shortcut. We want the secret to lose weight, make money, or get abs of steel, but we don't want to do the work it takes to get them. Hebrews 12:11 (ESV) says:

> *"For the moment all discipline seems painful rather than pleasant, but later it yields the peaceful fruit of righteousness to those who have been trained by it."*

Not only are those shortcuts all too often false, but even if they deliver on what they promise, they also bring deep dissatisfaction along too. There is no substitute for sweat and hard work.

TRAINING BEFORE WE NEED IT
It's also critical to realize that training must take place before it is needed. What kind of warrior doesn't put on their armor until they are being engaged by the enemy? The time to prepare is now. As the saying goes, "Dig a well before you're thirsty."

What specific things do you need to be preparing for now? These could be things like marriage, temptation, lust, fatherhood, financial stability, or anything that is ahead of you that you must to be ready for.

_____

_____

DEVELOPING YOUR NATURAL REACTION

When people say that something is a "natural reaction" they often mean that it couldn't be helped, or it's how everyone would have reacted in that situation. But our natural reactions are developed as muscle memory. Even an outburst of anger has been learned over time and allowed to be "natural." Or we use the old excuse of "that's what my dad or mom did" or "I don't know any better" which are just excuses to not change.

The bottom line is that even if everyone else genuinely does do it, we are not called to be like everyone else – we are called to be like Christ. Are there any "natural reactions" you are prone to that you shouldn't be? If so, write them down in the space below:

(Examples: Anger, Lust, Laziness, Pride, Temper, Avoidance)

_____

_____

_____

_____

DAY THREE
THE BEST INSTRUCTION MANUAL

In Jeff's teaching this week, he called the Bible "the greatest manual ever written" and one of the essential fundamentals in your Christian walk. Let's take a closer look at this manual and why it's so different.

A UNIQUE MANUAL
The Bible isn't comparable to any other book on Earth, and our attention to it should reflect that. Just look at a quick overview of this book that is all too often left collecting dust:

◆ Written over the span of more than 1,500 years on 3 continents and in 3 languages

◆ Written by more than 40 people, including Kings, peasants, musicians, politicians, philosophers, and fishermen.

◆ Written in the wilderness, in exile, in palaces, and in prisons.

◆ Written in joy, sorrow, despair, extreme clarity, and absolute confusion.

◆ Includes letters, poetry, history, songs, romance, political treatise, satire, biography, law, and prophecies.

◆ The entire Bible has been translated into more than 530 languages, with portions of it translated into more than 2,300 additional languages.

◆ It presents the story of God restoring the relationship between Himself and humanity through Jesus Christ, His Son.

Do you know anyone who is exceptionally disciplined in how they read their Bible? Who? How do they talk about God's Word, and what kinds of words do they use?

_____

_____

_____

WHAT IS STOPPING YOU?
You may find yourself coming up with all kinds of excuses of why you can't, don't have time, or are excused from reading your Bible today. The enemy wants to keep you from picking up this book, because he knows that God's word has power in it.

What excuses do you use when you don't want to spend time reading your Bible?

_____

_____

_____

YOUR ONGOING DISCIPLINE
As you continue to immerse yourself into God's Word, you will find strength when you need it, peace when you're worried, and clarity when you're confused. Even greater than that, you will find God's heart being shown to you, and you will find yourself being changed into the man God designed you to be. **Don't compromise that by being too proud to read the manual.**

Spend some time praying today that God would convict your heart and move you to action. Ask Him to give you a hunger for His word and to point you to something that will captivate your attention.

DAY FOUR
COMMUNICATION WITH OUR GOD

Prayer can be compared to getting on the radio and calling in air support. Communication through prayer is essential for Christians, not just for our good, but also for the good of those counting on us.

Prayer can be extraordinarily powerful. Just look at Exodus 32:9-14 - God decided to destroy all of Israel after their disobedience, but Moses prayed and God relented, sparing the entire nation! Prayer has the power to alter world events, like in this example, and it has the power to alter our hearts, bringing us in line with God's plan.

KNOWING HOW TO PRAY

For those who feel like they don't know how to pray, Jesus tells us how in Matthew 6:9-13 (NIV):

"Our Father in heaven, hallowed be your name,
your kingdom come, your will be done, on earth as it is in heaven.
Give us today our daily bread. And forgive us our debts,
as we also have forgiven our debtors.
And lead us not into temptation,
but deliver us from the evil one."

You will find, as you pray like Jesus, that you will learn the habit of seeing God as your Father, and that His name is unrivaled; that His rulership is coming to Earth and this is your deepest desire; that you should ask for what you need today; that you can forgive others because you are forgiven, and see that He is your protector and rescuer.

A WANDERING MIND

One of the most common struggles for men trying to have a consistent prayer life is that their mind tends to wander. Try some of these things to help limit how your mind wanders during prayer.

- ◆ Pray out loud or write down your prayers to help keep your train of thought together.

- ◆ Recognize that it is a spiritual battle – You are being distracted on purpose to keep you from praying.

- ◆ Praying scripture, especially the Psalms is a great way to help keep your heart on point.

- ◆ Go for a walk and pray outside. You may find that the motion of walking will quiet the to-do list in your head.

- ◆ Having the accountability of praying with someone else may also help structure what you pray about.

- ◆ If you are married pray with your spouse before you go to bed. Thank God for the things in your day. Pray out loud and don't be ashamed if you don't know exactly what to say.

What are some things that distract you while you're praying? What can you do to get rid of those distractions?

_____

_____

Remember, God wants to be communicated with – He has invited you to spend time with Him, and you will find that spending that time is always worth it.

At some point today, spend at least 10 minutes in prayer, trying one of the methods from the list above, maybe one you've never tried before. Then, when you meet with your group, talk about how that experience went.

DAY FIVE
HAVING A COACH

The third basic fundamental that Jeff mentioned this week was the need to have a coach – someone who has gone where you haven't yet and come back to tell you about it. No matter how old or mature you are, it is critical to have someone in your life who is further along than you, to help you get where you need to go.

STRENGTHENING THE RELATIONSHIP
Although you might have different coaches for different seasons of your life, you will find the longer you have one coach, the stronger that foundation becomes and the more impact they can have on your life.

In Hebrews 13:17 (ESV), it says:

> *"Obey your leaders and submit to them, for they are keeping watch over your souls, as those who will have to give an account. Let them do this with joy and not with groaning, for that would be of no advantage to you."*

Consider having a set time that you meet with your coach, and make sure he is free to ask you anything. This is a man who is invested in your success and who will be held accountable, so make sure you allow yourself to be honest with him.

IDENTIFYING YOUR CURRENT COACH
Maybe you already have someone functioning as your spiritual coach, but you don't call it by that term. It would most likely be the person you call to get advice on a big decision. Do some names come to mind? Write them down in the space below:

_____

_____

_____

HOW TO FIND ONE

If you don't currently have someone in that position, finding them may be easier than you think. Look around at the older people in your life and ask yourself which of them you would want to be like. Even if you already have a coach, ask yourself that same question and see if any additional names come to mind and write them below.

_____

_____

_____

BE A COACH

If you are at a place in your spiritual walk where you can help those coming along behind you, you need to be coaching someone else. God blesses us and helps us grow so that we can bless others and help them grow. Are there any men in your life who you can be coaching? Write down their names below:

_____

_____

_____

Today, touch base with one of the people you named. If they are already your coach, thank them for pouring into you. If you want to coach them or be coached by them, set up a time to meet with them soon. It is critical that you get a coach if you don't already have one. ***Missing out on that relationship could be an incredibly costly mistake.***

I AM NEVER
OUT OF THE FIGHT

DAY ONE
BEING SITUATIONALLY AWARE

Imagine having an enemy, dedicated to destroying you, who could attack you at any moment, even in your own home. In 1 Peter 5:8 (NIV), scripture tells us:

> *"Be alert and of sober mind. Your enemy the devil prowls around like a roaring lion looking for someone to devour."*

This means that no matter what, we are never out of the fight. We have an enemy that is always trying to destroy us, like a lion following us around everywhere we go, waiting for his chance to strike. We have to be on guard at all times.

WHAT DOES THIS LOOK LIKE?

We are in enemy territory and our enemy is always looking for our weak point. The good news is that God has equipped us with weapons for engaging our enemy, and they are effective against him. But we need to be ready to fight with those weapons at any moment.

In what ways does Satan try to attack you? How can you be "alert and of sober mind," being ready to face temptation to pornography, pride, anger, or jealousy at any moment?

_____

_____

_____

_____

YOUR PERSONALIZED WARFARE

Not only can our enemy attack us at any moment, but he has also studied us for years and knows the areas where we are most susceptible. He uses this intel to craft the most compelling temptations for each person. Just think about what Satan tempted Jesus with in Matthew 4:1-11:

- ◆ Perform a Miracle to Prove You are the Son of God
- ◆ Fulfill Old Testament Prophecy and Overcome Death
- ◆ Reign in Power Over All the Nations of the World

These were the very reasons Jesus came to Earth and Satan used them as bait. Jesus accomplished the first two goals and continues to accomplish the third, but in obedience to God, not as a shortcut. **Satan will offer you shortcuts to the things you want most in the world, but taking them will always destroy the things you love and value the most.**

COMPROMISING VIRTUE

Our enemy is very crafty and often engages us in battle the moment after we deliver him a blow. No sooner do we humble ourselves to serve someone else does the thought "Look at me, serving others! I hope someone notices how humble I'm being" enters into our heads.

Have you ever had an experience where you were attacked right after gaining victory? What temptation did the enemy use?

_____

_____

_____

_____

As you go through your day today, imagine that lion following you around, waiting for his chance to strike. If you are tempted to do something, imagine that lion poised to jump, depending on how you respond. Keeping that image will change your decisions and response time.

DAY TWO
KNOWING YOUR ENEMY

In special operations, it's key to know who your enemy is, what their capabilities are, and what weapons they have. This also remains true when it comes to our enemy, Satan. We need to know his tactics and techniques, so we can know how best to respond.

KNOWING WHO WE ARE UP AGAINST

The Bible is our best source for knowing the nature of our enemy. Think of the verses below as intelligence that has been gathered about your enemy. Look up each verse and write down key insights you gain about the enemy in the space below.

| James 4:7 | 2 Corinthians 11:14 | Hebrews 2:14 | Daniel 8:25 |

_____

_____

_____

_____

_____

_____

_____

SATAN THE DISTORTER

Satan has never had an original thought. His only ability is taking the good things that God creates and distorting them into perversions.

For example, God created sex to be enjoyable within marriage and Satan distorts it into all kinds of perversion. God has righteous anger against sin and Satan distorts this into anger over any perceived offense.

He is a master of lies, and knows the most effective lies have a bit of the truth in them. Write down areas that you are often tempted in and how Satan is distorting it. This could include things like: getting drunk or looking lustfully at a woman or greed - this is a distortion of God's intended design.

_____

_____

_____

_____

_____

_____

POWER TO OVERCOME

Spend some time in prayer this morning, asking for God's help in facing the enemy. Our God is unrivaled in His power, and you are His son. When you are under attack, call for help from Him, and the enemy will go running. Remember that 1 John 4:4 (ESV) says,

> *"...the one who is in you is greater than he who is in the world."*

DAY THREE
TACTICS OF DISTRACTION AND TEMPTATION

Jeff said that two of Satan's master tactics are temptation and distraction. He wants to make you ineffective by getting you to focus on less important things. Remember, both of these will happen to believers at all levels, but the most important thing is what you do after you face it.

MANIPULATING YOUR PERCEPTION
Distractions are often good things that take the place of more valuable things. God wants you to be a man who values the best things, leaving lesser things alone. **Satan doesn't mind you fighting, as long as it's a battle that doesn't matter.**

What are your distractions? Satan already knows what they are, and you don't want to be less honest than your enemy, so be specific. These could include TV, social media, certain relationships, or even your job.

_____

_____

BEWARE OF RATIONALIZATION
In temptation, the first red flag to look out for is rationalizing why sin isn't sin in this case. For example, "I wouldn't have lost my temper if I hadn't slept so poorly." Go on high alert when you start the mental acrobatics of rationalizing sin. **Truth never needs exceptions made for it.**

The thing you are convincing yourself isn't sin may be the thing holding you back from what God has in store for you. What things in your life are you rationalizing right now?

_____

_____

TACTICS FOR OVERCOMING

Ignoring temptation does not make it go away – you need to address it immediately. In 1 Corinthians 10:13 (NIV) it says,

> *"No temptation has overtaken you except what is common to mankind. And God is faithful; he will not let you be tempted beyond what you can bear. But when you are tempted, he will also provide a way out so that you can endure it."*

There is always a way out. Train your natural response when you are tempted to look for that way out and quickly take it. Below are some good immediate responses to help.

- ◆ **Call Out:** Pray out loud for strength or call a friend to pray.
- ◆ **Speak Out:** Read your Bible out loud to yourself.
- ◆ **Get Out:** Physically leave things that are temptations to you.
- ◆ **Cut Out:** Remove access to things that are temptations.

Make an action plan now, before you are tempted or distracted, and stick to it when you're in the moment.

WHAT ABOUT WHEN I FALL?

When you fall to temptation or distraction, Satan is going to try to keep you down with guilt. Instead, let the process of repentance and forgiveness push you closer to God rather than farther away. **The enemy might knock you down, but it's your choice to stay down.**

Spend at least 10 minutes today praying that God will help you respond quickly, keep your eyes on the most important things, and get back up when you're knocked down.

DAY FOUR
OUR OWN WORST ENEMY

It's true – we are in a war where there is a clear enemy. But the enemy who most often gets overlooked is ourselves. As fallen people, we have a sin nature built into us that we have to also do battle with on a daily basis. The Apostle Paul says in Romans 7:21-22 (ESV),

> *"So I find it to be a law that when I want to do right, evil lies close at hand. For I delight in the law of God, in my inner being, but I see in my members another law waging war against the law of my mind and making me captive to the law of sin that dwells in my members."*

There is a part of our hearts that is still being changed into God's likeness, and it does not want to change. It's not enough to focus on the enemy outside us – we need to also be on guard against our old nature.

THE PROCESS OF SANCTIFICATION

When you accept Christ, you are transformed into a new kind of being – this is salvation. You are also continually becoming more like Christ – this is sanctification. In terms of warfare, we have switched sides, but we're still figuring out what being on this side of the war looks like.

One of the most difficult things to learn after switching sides in the war is how to fight an old sin nature that used to be very familiar and comfortable.

In your Christian walk, what kinds of "old nature" things have you had to struggle with, or are currently struggling with? These could be anything from "sleeping with my girlfriend" to "bragging about myself."

_____

_____

JUST TRYING HARDER

It can be hard to be the one standing in your own way. Often, the immediate response is to strong arm that old nature out and try as hard as we can to be different.

The problem with this is that it never works for long. Sooner or later, our resolve fails and we are back to the old things we struggled with. **The only way to become more like Jesus is through submitting to the power of Jesus every day.**

REFINING GOLD

God uses many kinds of experiences to sanctify us, even difficult trials. The Bible compares being sanctified through trials to gold being refined in a fire. The heat burns off all the impurities in the gold, leaving only the pure behind.

What kind of experiences have you had where God molded your character through a difficult trial or experience? What was refined or burned away in that process?

_____

_____

_____

_____

_____

Spend a few minutes in prayer, asking God to help stand against the enemy of your own flesh and identify when it's getting in the way.

If something frustrating happens today, look for how God can use it to change you to be more like Christ. Even the frustrating things in your life can be used to make you more like Christ.

DAY FIVE
THE ARMOR OF GOD

In the military, soldiers are given equipment for warfare. These include helmets, body armor, belts, boots, and a weapon. In our spiritual walk, the Bible tells us in Ephesians 6:14-18 (NIV) that we are given equipment for a spiritual war, and that we must put it on every day.

*"Stand firm then, with the **belt of truth** buckled around your waist, with the **breastplate of righteousness** in place, and with your **feet fitted with the readiness** that comes from the gospel of peace."*

*In addition to all this, take up the **shield of faith,** with which you can extinguish all the flaming arrows of the evil one. Take the **helmet of salvation** and the **sword of the Spirit,** which is the word of God. And **pray in the Spirit** on all occasions with all kinds of prayers and requests."*

WHAT THIS LOOKS LIKE

This means every day you wrap the Truth of God around you like a belt so you are ready to move quickly to the fight. Protect your integrity like you protect your chest, and be ready with the Gospel on a moment's notice in any circumstances.

Every temptation will be ineffective if you protect yourself with Faith. Satan will try to make you doubt your salvation, so protect your head against attack.

Finally, have the Word of God out in front as your main weapon and stay in communication with God so you will be ready to go wherever He leads you.

SWORD OF THE SPIRIT

The Word of God is a unique part of this armor, since it is the only one that is both for defense and for offense. Knowing when and how to use it takes wisdom and discernment.

Who do you know who uses the Word effectively for both defense and offense? What do they do for defense and what do they do for offense?

_____

_____

_____

_____

BEING PREPARED

You can't use what you don't have, so get in the Word every day as preparation for when you will need to use that Truth to fight the enemy. It grieves God's heart when we ignore the tools He gave us, so get in the habit of putting on this armor every morning.

Are there things you need to modify in your morning routine so you will have time to put on this armor?

_____

_____

_____

_____

Find a verse you want to keep in mind today, and spend 10 minutes reading it over and over again, committing it to memory. Write it somewhere you can take with you and test yourself throughout the day. These small disciplines begin long term habits that transform your life.

LEAD AND BE LED

DAY ONE
YOUR SWIM BUDDY

From the beginning of training, each Navy SEAL is paired with a "Swim Buddy," someone who is always there in the blood and sweat of day-to-day life. Jeff talked this week about the importance of having this kind of relationship in your Christian walk. If you don't have someone in your life like this, it's time to find him.

THE IMPORTANCE OF HAVING ONE
In Proverbs 27:17 (NIV), it says "As iron sharpens iron, so one person sharpens another." God designed you to rely on and be strengthened by other people. **Isolation easily allows for sin to stick around,** but that swim buddy will call you out when you're slipping and stand guard when you're most vulnerable – and you will do the same for him.

Has there been a situation – maybe even this week – where having that swim buddy either made a difference or could have made a difference if it had been there? What was the situation?

_____

_____

_____

_____

_____

MUTUAL ADVANTAGES
There are many times when you will need to disciple or be discipled by someone else, but the relationship with your swim buddy is one where you are both continually sharpened by each other.

In Ecclesiastes 4:9-12 (NIV), it says,

"Two are better than one, because they have a good return for their labor: If either of them falls down, one can help the other up. But pity anyone who falls and has no one to help them up. Also, if two lie down together, they will keep warm. But how can one keep warm alone? Though one may be overpowered, two can defend themselves. A cord of three strands is not quickly broken."

This relationship should be helpful for both people – not just one. This forces you to think in terms of how you work as a member of a team, rather than just for your own advantage.

WHO IS YOUR SWIM BUDDY?

This is a very deep relationship where you both spend a lot of time together, but your swim buddy is not your spouse. You need the accountability of preparing for battle and being in battle with other men. Do you have a swim buddy? If you do, write down their name below. If you don't, write down some names of people who could be yours, but aren't yet. Keep in mind that some swim buddies can be for just a season of life that you are in.

_____

_____

_____

If you already have one, take a few minutes and pray for him today. You are invested in his success, so reach out to him today and find a way to encourage him.

If the names you wrote down are of potential swim buddies, spend some time right now praying for each name you wrote down. Ask God to reveal who you should ask.

DAY TWO
YOUR RESPONSIBILITY IN COMMUNITY

As men, we often fall into glorifying the loner hero. From John Wayne to John McClane, we think that strength means handling everything by yourself. In the Navy SEALs and in the Kingdom of God, there is a very different idea of what strength looks like.

TRUE STRENGTH
In Romans 12:4-5 (NIV) scripture says,

> *"For just as each of us has one body with many members, and these members do not all have the same function, so in Christ we, though many, form one body, and each member belongs to all the others."*

As a Christian, you are called to be in community with other believers. Strength, under these terms, means working with and not against the other members of the body of Christ.

Even Jesus, who was God on Earth, submitted to the authority of the Father and called believers to be united. Note that when He taught his disciples to pray, He didn't say "My Father in Heaven" – He said "Our Father in Heaven."

So the question isn't if you have community – the question is what are you doing with the community you have? Are you being real with your group, or do you stay emotionally disengaged and compartmentalized? Write down who is in your current community and how you can engage with them. If you don't have a group of 2 or 3 men that you connect with regularly then you have to start there. Maybe they are in your Common Man | Uncommon Life group.

PULLING YOUR WEIGHT

A team member who doesn't contribute or pull his weight becomes a liability rather than an asset. Isolating yourself from your community has serious consequences, not only for you, but also for those around you.

Each person has been gifted with certain talents and skills. When these skills are combined, they allow a group to be stronger than the sum of their parts.

Write down someone in your group who has a strength you don't have and what that strength is.

_____

_____

_____

RECOGNIZING YOUR STRENGTH

It's critical to also know how you can invest in those around you. Think of a skill or discipline that comes easily for you, but other people say it's hard for them. Write it down below:

_____

_____

Spend at least 10 minutes today praying for the people in your community. Ask God to show you ways you can invest in them even more and how you can learn from their skills in your daily life. Ask God to help you be honest and open with them.

DAY THREE
BEING UNDER AUTHORITY

Jeff talked about two types of male relationships in the video this week – a peer relationship (swim buddy), and an authority relationship. The Bible has a lot to say about being under authority, to both God and your leaders, but there are some challenges that can quickly come up.

UNDER AUTHORITY TO YOUR LEADERS
The idea of submitting to another person and being under their authority can be very hard for many men. It can be uncomfortable letting someone else have the power to tell you what you should and shouldn't be doing, but it's an essential part of growing as a believer. In 1 Peter 5:5 (ESV), it says:

"For just as each of us has one body with many members, and these members do not all have the same function, so in Christ we, though many, form one body, and each member belongs to all the others."

The "younger" and "elders" here applies to age as well as spiritual maturity. No matter how old or mature you are, there should always be men in your life who are further along, looking out for you.

Write down the names of people who you look to as an authority in your life. They could be father figures, pastors, more mature believers,

_____

_____

_____

_____

WHAT STOPS US

The second part of this verse gives us the key to why obedience to authority is so hard – it's because of our pride.

Pride tells us to trust in ourselves and that submitting to others is weakness. **The problem with pride is that it makes us blind to our sin, and we will keep running into the same issues, over and over again.** Proverbs 16:18 (NIV) says "Pride goes before destruction, and a haughty spirit before a fall."

Are there authorities in your life that you are currently not submitting to? What can you do differently this week to honor them?

_____

_____

_____

UNDER AUTHORITY TO GOD

As a Christian, you are primarily a member in the Kingdom of God, and obeying the King comes with the territory.

Fortunately, our King has always and will always have our ultimate best in mind. Obeying Him and submitting to Him will always bring you where you need to be. It might not be easy, but it will always be worth it.

Spend a few minutes asking God to search your heart and show you any areas you are disobeying in. If there are, write them down below. What can you do today to be obedient?

_____

_____

_____

DAY FOUR
THE LEADERS YOU SHOULDN'T HAVE

Yesterday, you looked at the people in your life who you look up to as leaders. It is just as important to recognize the people and things that shouldn't be in authority in your life. Proverbs 13:20 (ESV) says,

> *"Whoever walks with the wise becomes wise, but the companion of fools will suffer harm."*

You will always become like the people you spend time with and look up to. Choosing who these companions are is vital to your emotional and spiritual health.

IDENTIFYING BAD INFLUENCES

Friendship can run very deep and recognizing the peers and leaders who are a bad influence on you can be difficult. Take some time and think through the close friends you have and ask yourself these questions:

- ◆ Do you fall into bad or old habits when you're around them?
- ◆ Do they share and support your goal of becoming like Christ?
- ◆ Do they show wisdom in how they make decisions?
- ◆ Are they honest with you, even when it's uncomfortable?
- ◆ Can you talk to them about struggles you're having?
- ◆ Do they give you good advice?

After asking yourself these questions, are there people who have influence in your life who shouldn't? Write down any names that come to mind:

SUBTLE INFLUENCES

Not all leaders are people. Every day, there are things in your life that influence you to act or not to act in a certain way. Read through the statements below and see if any of them sound like reasons you use to do or not do certain things in your life.

- It's what I've always done
- Everyone does it that way
- I would be weird if I don't/do
- I might make someone uncomfortable
- I don't want to
- It's hereditary

These kinds of appeals can harbor addictions like pornography and alcohol abuse, control how we spend our time and money, and what we choose to say or not say. That kind of power should only be given to things that make us more like Christ.

What influences in your life do you need to take control away from and what steps, maybe even radical steps, can you take to alleviate and truly make a change?

_____

_____

_____

Take 10 minutes and pray over the list of people or influences that you wrote down. Ask God to show you areas where you are giving control to people or things you shouldn't and what your next steps are to give control back to Him. Don't be afraid to share these with your swim buddy or someone you trust to hold you accountable. This may be the first step in a new journey.

DAY FIVE
BEING IN AUTHORITY

Whether or not you consider yourself a leader, there will always be areas in your life where you are held responsible. From how you lead your family to how you influence those around you, there are things in your life that God has given you authority over.

BEING WORTHY OF HONOR
In Proverbs 29:2 (ESV), the Bible says "When the righteous increase, the people rejoice, But when a wicked man rules, people groan." Think about a leader you have who is worthy of respect. Write down some of the character qualities that contribute to why you respect them.

_____

_____

_____

When you lead others, do you show these character qualities to those you're leading?" If not, what is standing in your way?

_____

_____

_____

_____

POWER AND AUTHORITY
In leadership, it's important to understand the difference between power and authority. Power is your **ability to act** and authority is your **right to act**. On the other page are some profiles of leaders who use power and authority in different proportions:

- **The Powerless Leader** has authority to act, doesn't have or doesn't use their power to do so. Example: A boss who doesn't engage with and lead their team at work.
- **The Tyrant Leader** has power and authority, but lets power determine their control, rather than authority. Example: A father who over-disciplines his own children.
- **The Bully** exercises power over things that are not under their authority. Example: A parent trying to discipline someone else's child.
- **The True Leader** lets their authority set the boundaries for how their power is exercised. Example: Jesus submitting himself to His Father in being human.

Write down any relationships you are currently in where your ability to act and your right to act are out of proportion. For example, sleeping with someone you are not married to, being emotionally disengaged from your family, or abusive family or work relationships.

_____

_____

_____

_____

_____

YOUR SPHERE OF INFLUENCE

In the video this week, Jeff asked you to write down names of people you are in authority over in your work, church, and community. You are not in these relationships accidentally. God has given you this influence for a reason, and everyone is held accountable for what they do with what God gives them.

Spend some time praying over the names on your list. Make it a habit to pray for these people. What can you do – today – to invest your influence for the Kingdom?

MY CHARACTER AND HONOR
ARE STEADFAST

DAY ONE
YOUR CHARACTER

In the video this week, Jeff defined character as "The place inside you where the things you say and the things you do and the decisions you make come from." Jesus talked about this in Luke 6:43-45 (NIV):

"No good tree bears bad fruit, nor does a bad tree bear good fruit. Each tree is recognized by its own fruit. People do not pick figs from thornbushes, or grapes from briers. A good man brings good things out of the good stored up in his heart, and an evil man brings evil things out of the evil stored up in his heart."

In other words, your character is the source of your actions – it is the reason you do what you do, for good or for bad. This verse tells us that we can recognize both good and bad character by the end result.

RECOGNIZING YOUR CHARACTER

We cannot disconnect how we act from who we are. This means that simply excusing anger as an outburst, lust as normal, or getting drunk as an occasional indulgence, is not an option. Our decisions clue us in to who we are inside, and true character is defined by who you are when no one else is watching.

Look over the following questions and write about at least one of them in the space on the next page. What can you learn about your character by how you respond to these situations?

- How do you spend your money?
- How do you spend your free time?
- How do you treat people you don't like?
- How do you use influence or power when you have it?

_____

_____

_____

_____

_____

TRANSFORMATION AT THE SOURCE

How you behave is the result of who you are, but changing who you are must go deeper than simply modifying your behavior. Changing your character by changing your actions is like trying to improve the speed of your car by adding racing stripes. It may change the appearance, but not the internal workings.

Jesus came to change who you are, not just how you act. But you will notice that as you become more like Christ, your actions can't help but change too.

Think about times in your life when you tried to change your behavior in an effort to change your character. Did the change last? What was the outcome?

_____

_____

_____

As you go through your day today, ask yourself the question, "What do my thoughts and actions tell me about my character?" As you see the decisions you make and things you do, pray that God would reveal areas where your character is weak and that He would give you strength to become more like Jesus.

DAY TWO
A MAN OF INTEGRITY

Character is the source of your actions and integrity is the consistency of your actions. Having integrity means that you are the same man in public and when nobody is watching. In Proverbs 10:9 (ESV), the Bible says:

> *"Whoever walks in integrity walks securely, but he who makes his ways crooked will be found out."*

Lacking integrity is like a tree that appears strong and sturdy on the outside, but when it's under pressure, it breaks because the inside was not as strong as it appeared.

CONSISTENCY IN WHO YOU ARE

When you know the strength you project on the outside does not match the condition of your heart, there is a fear of being found out, of being tested, and of being relied on.

Are there areas of your life where you are living different values, based on who is around? If so, write down any areas that need attention below.

_____

_____

_____

_____

Satan will tell you lies about these areas to keep you from acting. He uses fear of being found, shame from past failures, and feelings of unworthiness to convince you to hide your sin. Be the kind of man who fights these lies and runs toward character and integrity.

A MAN OF YOUR WORD

Following through on your promises is good evidence of integrity in your life. **A man of integrity does not take his commitments lightly and sticks to what he says he will do.** When those around you know you are a man of integrity, they give more weight to your word, because they know you will follow through.

Look over the statements below and ask yourself if they are true of you. Would the people close to you say the same thing?

- ◆ When I tell someone I will pray for them, I do.
- ◆ When I make a commitment, I follow through.
- ◆ People can count on me to be there when I said I would be.
- ◆ I take responsibility for my mistakes.

Are there areas of integrity that need your attention? If so, write them down below and tell your group to help keep you accountable.

_____

_____

WHAT IS AT STAKE?

There is far more at stake than just your personal reputation. Integrity is your foundation – ensuring it is sound means you are able to build on it and become the man God is calling you to be.

Without it, you are putting not only yourself in danger, but also the people around you. **The people close to you count on you to be the same outside as you are inside.**

Right now, spend some time asking God to show you any areas of inconsistency in your life. Picture these areas as cracks in a foundation, and ask God to inspect your character and help you become a man of solid integrity.

DAY THREE
REFUSING TO COMPROMISE

Jeff told the story of Shadrach Meshach and Abednego, and their refusal to compromise, even at the threat of death. In our day-to-day lives, compromise often comes in more subtle forms than life or death scenarios, but they are no less important.

Your small battles of compromise are building or breaking you. You have to decide what things you aren't willing to compromise long before a test as serious as the one those three men faced.

MAKE NO PROVISION

Tests of your character are going to come and preparation can make or break you. In Romans 13:13-14 (NIV), it says,

"Let us behave properly as in the day, not in carousing and drunkenness, not in sexual promiscuity and sensuality, not in strife and jealousy. But put on the Lord Jesus Christ, and make no provision for the flesh in regard to its lusts."

When the temptation to compromise comes, make sure you "make no provision" or protect a way for you to give in to it – Cut off that road ahead of time.

This could look like removing alcohol in your house, having an Internet filter on your computer, or giving oversight of the finances to your spouse, or putting a check in your responses before you speak them.

Are there sins in your life that you are making provisions for? What steps can you take to limit your options when you are tested?

SMALL ALTERATIONS

The enemy, especially in times of testing, will tell you that compromise is reasonable, isn't a big deal, and it's not like you're as bad as that other guy.

But just like your character is built over time by being faithful in small things, large compromises are also developed one small step at a time. There is no reasonable level of compromise when it comes to sin. Satan's goal is to make a crack and widen it as far as he can, and to do that, he always starts small.

Write down areas in your life that you are either prone to compromise in or are currently compromising in, as well as lies that the enemy tells you in those times of testing.

_____

_____

_____

_____

Take some time and pray over the list above. Ask God to help you be ready for testing in these areas, and as soon as you hear those lies from the enemy, go on high alert and refuse to give ground.

Remember, your enemy is not interested in negotiating – he is only interested in destroying. You cannot afford to compromise to any degree.

DAY FOUR
WORSHIP MATTERS

In the story of Shadrach Meshach and Abednego, their faith was tested when the king ordered everyone to bow down to a statue, and anyone who did not would die.

Those men recognized that worship is essential to Christian character, and that man was created to only worship God. Daniel 3:17-18 (NIV) says they responded by saying,

"If we are thrown into the blazing furnace, the God we serve is able to deliver us from it, and he will deliver us from Your Majesty's hand. But even if he does not, we want you to know, Your Majesty, that we will not serve your gods or worship the image of gold you have set up."

Refusing to worship anything less than God and identifying things that might be taking His place is a critical part of building and protecting your character and integrity.

WHAT YOU VALUE
Below, make a list of the top things, people, and activities you value the most. Group similar things together, and don't aim for just good Christian answers – be honest and really evaluate what you highly value, for good or bad. Write down at least 10.

_____

_____

_____

_____

_____

Now write down the things you do the most in any given week. These could include things like work, sleeping, watching TV, reading, etc. Write down at least 10 here too.

_____

_____

_____

Remember that our enemy is crafty. He knows worship has to be subtler than bowing down to idols. He also knows that the shortest distance to 1st place is whatever is in 2nd place or something you are already used to doing often.

Look over both lists and ask yourself if anything on there that could be taking the place of the authority, honor, and value you should be giving only to God. A good indicator of what you highly value is what gets you upset.

Spend a few minutes really looking over the lists and weighing each one. The danger of putting something in God's place is something Christians at all levels of maturity need to guard against.

_____

_____

Take a couple minutes and pray the following:

"Heavenly Father, I want to use my time, money, and emotions today on what matters most to you. Please show me any areas of my life that are competing for my worship of you, and give me the strength to act boldly on what I see. Thank you for being patient with me and for your unfailing love. Amen."

DAY FIVE
CONTINUALLY BUILDING YOUR CHARACTER

Building and protecting your character is not something that you can afford to do once and then never look at again. Jeff described being a Christian as an ongoing lifestyle – something that is deeply rooted in every single day. Colossians 2:6-8 (NIV) says,

> *"So then, just as you received Christ Jesus as Lord, continue to live in him, rooted and built up in him, strengthened in the faith as you were taught, and overflowing with thankfulness."*

PREPARING EACH DAY

Being rooted and built up does not happen instantly. It is something you fight for every day and continue to deepen and strengthen. **As you spend time in God's word and in prayer, He will show you areas in your character to fortify, making you more like Christ.**

Your internal character is one of the most valuable things in your charge. How often do you set aside time in your week to read God's word and pray, asking Him to help you strengthen your character? What things are standing in the way of seeking Him more?

_____

_____

_____

_____

KNOWING AND ACTING

As God continues to show you areas in your life to surrender to Him, you must be faithful in following through. It's not enough to talk a big game – you need to be the kind of man who acts on what he knows.

Being a man of character and action means that you see every situation as an opportunity for God to use. From a flat tire to being passed over for a promotion, recognize that God can use any frustration or trial to mold you, if only you'll let Him.

Think back to a recent frustration you experienced. Are there things you could have done differently to let God build and strengthen your character?

_____

_____

_____

_____

BEING PREPARED FOR SOMETHING

As you root out addiction, anger, insecurity, and pride, remember that the struggle is forging you into a certain kind of man – the man God has called you to be. It is a slow and steady process.

The same principle is true for habits of sin. As you protect areas of sin in your life, they are also preparing you to be a certain kind of man – but that man is weak, and selfish, and cowardly. **Every day you are choosing the kind of man you will become.**

Take a couple minutes and pray about how you should finish this sentence:

"I want to be the kind of man who_____."

Then keep that phrase with you today, and ask yourself if what you are doing today is helping you become more like that man. Ask God to continue to make you like that man.

I HUMBLY SERVE

DAY ONE
DO SOMETHING AT YOUR EXPENSE

Jeff mentioned five rules he gives to the young men he mentors. You can learn more about the other four rules at www.lightthedark.org, but he mentioned that rule #3 is "Do something that benefits someone else at your expense."

SERVING OTHERS
Every believer has been called to be a servant of others. Jesus set this example for us in John 13:13-16 (NIV), when he washed his disciple's feet. He explained why He did this by saying,

"Now that I, your Lord and Teacher, have washed your feet, you also should wash one another's feet. I have set you an example that you should do as I have done for you. Very truly I tell you, no servant is greater than his master."

If you think that you are too good to serve others, you are saying that you are better than Jesus – because He was the best man who ever lived, and even He lowered himself to serve others.

HOW ARE YOU SERVING?
As Christians, the question should never be if we are serving, but in what way we are serving. Think through the areas of your life – work, church, community, family, etc., and write down ways that you are serving the people in your life.

_____

_____

_____

_____

UNDERSTANDING "EXPENSE"

The second part of Jeff's rule is "...at your expense." This expense can come in many different forms, but all of them reinforce the importance of others rather than the importance of you.

What God has given you needs to be reinvested into serving others. Think about the areas of your money, time, reputation, energy, emotion, and position. How can you use these resources to serve the people you encounter on a weekly basis?

_____

_____

_____

_____

_____

_____

HOW TO SERVE
When you serve others, don't do it for recognition, appreciation, or payback. Be the kind of man who just serves others as naturally as water flows downhill. As you get lower to serve others, you will find that is exactly where Jesus has been waiting for you.

Remember, the Son of God lowered Himself down to be a servant. We are left without excuse for not serving too.

From your spouse to your coworker, think about the people you encounter each day, and ask God to show you ways that you can serve them today. As needs come up today, act quickly and serve right away.

DAY TWO
HUMBLY SERVING

In your goal to be the kind of man who humbly serves, one of the biggest barriers is pride. This may be one of the easiest sins to accept, but it is also one of the most damaging to our personal and spiritual lives.

1 Peter 5:5-6 (NIV) says,

> *"All of you, clothe yourselves with humility toward one another, because, 'God opposes the proud but shows favor to the humble.' Humble yourselves, therefore, under God's mighty hand, that he may lift you up in due time."*

Because pride is something that makes God oppose you, it is something that needs to be very clearly understood, so you can avoid it at all cost.

SNAPSHOT OF PRIDE

One of the marked characteristics of pride is that it puts you at the center of what is most important. You then view the world through your needs, desires, time, and possessions. At its core, pride is our internal desire to be powerful, in control, on top – to be God.

It also impairs our judgment of others and ourselves. It makes us believe we are stronger, wiser, and more capable than we actually are, and refuses to let us hear otherwise.

Think of a situation where being prideful got you into trouble. What was the outcome?

_____

_____

_____

TRUE HUMILITY

Humility often gets mistaken for self-deprecation or poor self-esteem, but it isn't either. C. S. Lewis once said **"True humility is not thinking less of yourself; it is thinking of yourself less."** In other words, humility always views others as more important.

This is the kind of heart that God desires His men to have, and humility is the foundation of serving others. Getting low and treating others as more important forces out every ounce of pride.

Who do you know that displays this kind of humility? Write about a time when you saw them putting someone else's needs in front of their own.

_____

_____

_____

One of the most effective ways to damage pride and build humility in your life is to make serving others a priority in your daily life. Because they are opposites, they can't coexist – one will always push the other out.

Are there people in your life who you find hard to serve because of your pride? Write down their names and ask God to show you one way you can serve them this week.

_____

_____

As you go through your day today, look for evidence of pride in your heart. Ask God to reveal pride in your heart and actions, and to show you ways you've acted in pride towards others.

DAY THREE
HEARING THE VOICE OF GOD

One of the most important things that any Christian can do is learn to hear and obey the voice of God. It is rarely a booming voice that comes out of heaven, but there are ways you can learn to listen for His voice.

In 1 Kings 19, the prophet Elijah has an encounter with God that helps us understand what God's voice is like. As Elijah waited for God, a powerful wind tore through, an earthquake shook, and a fire blazed, but it was only when he heard a gentle whisper that he got up to meet God.

Often, God's voice is a small prompting inside you or a thought you can't seem to shake. Does this way of thinking of God speaking sound like how you hear Him? If not, what are some other ways that God speaks to you?

_____

_____

_____

_____

_____

WHY GOD SPEAKS
When God speaks, it almost always ends in the area of serving others. It can be as simple as "maybe I should pay for that person's coffee" or "I should stop and see if they need help," but it doesn't end there. These acts of kindness grow into things like working day after day in a job you hate to provide for your family, or being there for your neighbor through a hard time or divorce.

When those moments like that do come, don't dismiss or mute it. Satan doesn't tempt people to serve, and he wants you to think about the cost to you, not the benefit to others. Don't get used to ignoring people rather than serving them.

God may be speaking, but are you making the time during your day to listen? When you do hear Him, do you quickly obey what He tells you? Write down how you can make time to listen today.

___

___

___

___

CUTTING OUT DISTRACTIONS

Think about things that may be getting in the way of hearing God's voice. Maybe you need to turn the radio off on your morning commute, or use your lunch break to spend a few minutes quietly listening. Write down at least one step you can take to make yourself more attentive to God's voice this week.

___

___

Be on the lookout today for God at work. When you pay attention and look for God, you will be surprised how involved He is in everything. When something unexpected happens, ask God "What do you want me to see here? What are you doing and how can you use me?" Then when He presses something on your heart, act right away.

DAY FOUR
LOVING THE UNLOVABLE

Jeff said that the Bible comes down to the four words "Love God, Love People," and that we are called to love all people, not just the ones we want to. But what does loving everyone look like in your day-to-day life?

GOD'S STANDARD OF LOVE
In Matthew 5: 43,44, & 46 (NIV), Jesus tell His disciples,

"You have heard that it was said, 'Love your neighbor and hate your enemy.' But I tell you, love your enemies and pray for those who persecute you…If you love those who love you, what reward will you get? Are not even the tax collectors doing that?"

The love of God is different from the world's kind of love because it is given to everyone, whether they deserve it or not. All of us have been on the receiving end of His love, and know we did not earn it. God's standard is this: **Give to others what God gave to you.**

PUTTING IT INTO PRACTICE
This is an easy thing to say, but it is much harder to do. Showing love to a person who frustrates you, slanders you, smells bad, or takes advantage of you is impossible without the grace of God. You need to accept His love as an unearned gift before you can give it to someone else on the same terms.

In the space below, write down the names of the top three people you interact with who really get under your skin and frustrate you.

1)_____

2)_____

3)_____

The first step in loving people who are hard to love is to start praying for them. **You will find that as you pray for them, you will start to become invested in their development.**

SEEING THROUGH GOD'S EYES

Loving people who frustrate you means you genuinely want the best for them, despite how they treat you. This perspective can only be achieved by seeing people how God sees them.

God sees that everyone is wounded and in need of healing, and those traits that frustrate you is how they are limping. Everyone has pain they are dealing with, and the cure for that pain is the love of God, which you have been given.

Look over the three names you wrote down, and now write down some ways you can be praying for each person. As you pray, wait on God and listen for more ways you can show them love.

_____

_____

_____

_____

LIKING VS. LOVING

Loving everyone is not the same as liking them, but you may find your attitude being changed as you pray for those people in your life. **God designed His love to change the person giving and the person receiving it at the same time.**

After praying for those people you wrote down, think of at least one way you can show love to them the next time you you meet them. Have those things locked and loaded and be on the lookout for when you can act.

DAY FIVE
WHAT KEEPS YOU FROM SERVING?

When face-to-face with people God has called you to serve, there can be many things that get in your way. Some may be deep hurts that God is healing you of over time, but often they are empty excuses that need to be identified and called out for what they are – excuses.

IDENTIFYING EXCUSES
Below are some common excuses that people make to avoid serving others. Read through the excuses and their answers, and circle any excuses you have found yourself making.

EXCUSE	ANSWER
It's easier to not.	You have not been called to take the easy route.
It would be uncomfortable.	If comfort is your goal, you're in the wrong game.
I don't have time.	Make time for what is important to God.
It would be hard.	Of course it's hard – things worth doing are always hard.

GOD'S ANSWER TO EXCUSES

When the prophet Jeremiah was called by God to speak, he told God why he couldn't do it, in Jeremiah 1:6-7 (ESV).

"Then I said, "Ah, Lord God! Behold, I do not know how to speak, for I am only a youth." But the Lord said to me, "Do not say, 'I am only a youth'; for to all to whom I send you, you shall go, and whatever I command you, you shall speak."

When God calls you to act, there is no excuse that can stand against Him. **God not only calls those who are equipped, He also equips those He has called.** As you obey what God tells you to do, you will find that it may be outside of your comfort zone, but it is not outside your ability.

What is God putting on your heart to do that you have been putting off? What does complete obedience to this look like, and what can you do today to obey?

_____

_____

_____

_____

EXPECT RESISTANCE

The last thing Satan wants you to do is serve others. As you obey God's call to serve, expect to meet resistance. But remember that strength is formed by pushing against resistance. Don't get discouraged – it means you're hitting him where it hurts.

Spend 10 minutes in prayer today, asking God to show you where you are able to serve, and to be prepared for resistance as it comes. When He shows you a way you can serve someone else, do it right away. Don't get in the habit of talking yourself out of serving.

I WILL GET BACK UP, EVERY TIME

DAY ONE
WHEN YOU FAIL

No matter who you are, a time will come when you will fail. You are a human, and getting knocked on your back will happen from time to time. **But the true test of a man is not whether or not he fails, but whether or not he keeps getting back up.**

Failing will turn your focus in one of two directions. You can keep looking at yourself or you can look to God. In Psalm 73:23 (ESV) it says,

> *"My flesh and my heart may fail, but God is the strength of my heart and my portion forever."*

Recognize the failure for what it is, learn from it, but the conclusion needs to be focused on God's grace, not on your failure.

Think of a recent time when you failed. What were some lessons you learned from it that you can take with you for next time. What are some ways you can turn that failure into focus, praise, or dependency on God?

_____

_____

_____

_____

DON'T WITHDRAW
Failure can be embarrassing and uncomfortable to talk about. We try to project a front of strength and power, so when we fail, the tendency is to hide it from other people.

This hiding after failure goes all the way back to Genesis 3, and it never works. Satan lies and tells you that you need to hide your sin, but darkness is sin's natural environment, so what you try to hide, you end up planting and watering instead.

Are there areas of failure or sin you are hiding and protecting from others right now? What is it going to take to break that stronghold in your life? This is a hard question, but it's one that needs to be answered. Take some time and ask God to help you see them clearly.

_____

_____

_____

_____

PEOPLE ARE COUNTING ON YOU
Whether you see it or not, there are people depending on you to get back up. **Part of your body can't just quit without affecting the rest of the body.**

Satan wants you to be ineffective and stay out of the fight when you get knocked down, but his end goal is to cripple the body of Christ. Do not let him. Get back up and go in for another round.

GOD'S GRACE
Don't sit and wallow in guilt over your failure – Jesus paid for every sin you ever committed as well as any sin you will commit. Don't use that grace as a license to sin, but instead, let it push you closer to God because of His goodness.

As God shows you areas where you are trying to hide sin, write them down and share them with your swim buddy. Think of ways you can do damage to those areas of sin in how you act today.

DAY TWO
MEN STRUGGLE, BUT THEY STRIVE

No one is going to find everything easy. Every man you will ever know is going to struggle with something and will have their own set of weaknesses. It is a false image to think that a real man finds everything easy or has no struggles.

Knowing it's not going to be easy means you need to be that much more prepared to fight for what is important. 2 Timothy 2:15 (ESV) says,

"Do your best to present yourself to God as one approved, a worker who has no need to be ashamed, rightly handling the word of truth."

God knows it won't be easy, but He has given you all the tools you need to fight well.

CORRECTING OUR VIEW
Many men try to live up to an impossible standard and put up a false front in the process. Be the kind of man who is secure enough in His identity in God to admit when you are having trouble and rely on the team He has given you.

What areas are you struggling with right now? Write down repeat offenders that give you trouble.

_____

_____

_____

_____

STRIVING FOR SOMETHING

The word "strive" has in it not only the idea of action, but also the idea of direction. You don't just fight against the things you wrote down – you are also fighting toward something else at the same time.

One good exercise is to fix your eyes on the virtue that the sin is fighting against and run toward that virtue.

When We Strive **Against**	We Are Striving **For**
Pride	Humility
Lust	Love
Anger	Peace/Self-Control/Gentleness
Resentment	Joy/Goodness
Unforgiveness	Forgiveness
Impatience	Patience

Spend 10 minutes in prayer, asking God to help you strive for goodness as you strive against sin. Look over the list of areas you wrote down and write down the virtue that those struggles are fighting against.

_____

_____

Keep the list of virtues with you today and look for situations where you can put them into practice. You might be surprised how many there are.

DAY THREE
REMOVING YOUR OPTIONS

In the video this week, Jeff said in your struggle with sin, you should limit your options and cut out the things that are causing you to stumble. In Matthew 5:29-30 (ESV), Jesus uses a graphic example to show how seriously he takes this principle:

"If your right eye causes you to sin, tear it out and throw it away. For it is better that you lose one of your members than that your whole body be thrown into hell. And if your right hand causes you to sin, cut it off and throw it away. For it is better that you lose one of your members than that your whole body go into hell."

The metaphor helps us understand the importance of limiting what we let ourselves do, when that freedom keeps tripping us up.

WHAT THIS LOOKS LIKE
If you struggle with pornography, you could set an Internet filter or timer on your computer to limit the content and time of day you can spend online. If you struggle with gambling, it could mean opting out of poker night with your friends.

Write down some temptations that you tend to struggle with and some options you can remove to help overcome it. Think of the path you would have to take to follow through on that sin and how you can put blockades on that path.

_____

_____

_____

SENDING A CLEAR MESSAGE
One of the benefits of doing this is that you set a standard for your future self to stick to. When you are faced with temptation, it is powerful to remember that there is a part of your heart that really doesn't want you to give in, and took steps to help you resist.

Many men would rather live with habitual sin than cut them off at the source. Maybe it feels too radical to not have the freedom of unlimited access to your TV, Internet, or finances. But radical change requires radical action.

WHAT IS STOPPING YOU?
Look over the list of options you wrote down and ask yourself, "What is stopping me from acting on these today?" Write down any excuses you come up with and recognize them as lies. You can make a drastic change, even today.

_____

_____

_____

_____

Don't delay in acting. **The best time to make plans for when you won't be thinking clearly is when you are clear headed.** You cannot afford to lose momentum and accept a pattern of sin as normal. Make sure you have a plan in your mind of how you will cross that bridge long before you come to it. SEALs train their muscle memory long before they need to be in battle.

Spend some time in prayer asking God for the strength you need to make hard decisions about how to guard yourself against sin. Then get up and make that change right now. Cut out the things in your life that aren't bearing fruit and make room for habits that will actually produce.

DAY FOUR
GOD'S POWER IN OUR WEAKNESS

It can be difficult to feel as though God is unable to use you because of a certain weakness in your life. But that's like saying you can't go to the gym because you're not strong. The good news is that God loves using us in our weaknesses, and has done so many times before.

Even Paul had a certain weakness. In fact, in 1 Corinthians 12:9-10 (ESV), he talks about how he begged God to take away "a thorn", but God had other plans.

"But he said to me, "My grace is sufficient for you, for my power is made perfect in weakness." Therefore I will boast all the more gladly of my weaknesses, so that the power of Christ may rest upon me. For the sake of Christ, then, I am content with weaknesses, insults, hardships, persecutions, and calamities. For when I am weak, then I am strong."

WHERE WE ARE WEAK
Paul shows us some very important things about God in these verses. For example, we can't use our weaknesses as excuses not to obey. Being a bad public speaker is not a reason to disobey God when He calls you to speak. **God often calls people to do things they could not do without His help.**

Areas we are not gifted in are prime spots for God to move in and show off His ability through you. What are some areas that you are weak in? These could be anything from "I'm not good at empathizing" to "I'm disorganized."

_____

_____

_____

WHY GOD USES US

God is a very creative investor. He calls us to use what we have to further the Kingdom, but many of us don't realize that our weaknesses are something we have that can be used even more effectively than our strengths.

When God accomplishes something great in someone who could not have done it on their own, the only explanation can be the power of God. This makes sure that people can see past us to see how God is using His people to change the world.

HOW GOD CAN USE YOU

Look at the areas you wrote down that you are weak in. Spend a few minutes praying, giving these areas to God for Him to use. Maybe there are already ways God has called you to act, but you have been disobeying. Write down some ways you can surrender your weaknesses to God's use and let Him receive glory in how He uses you.

_____

_____

_____

_____

_____

_____

_____

Follow up on this list with your swim buddy and see if there are any additional ways they see God using you in your weaknesses. As more come up in your conversation, add them to the list above.

DAY FIVE
FINAL RECAP

Thank you for participating in Common Man | Uncommon Life. We hope you have been affected by the content and that God used it to help draw you closer to Him.

Over the past 7 weeks, you have processed a lot of material, so this final day will allow you to reflect on the content you have finished and review some of your notes.

SUMMARY PAGE
On the opposite page, there is a summary sheet of things to fill out. Once you fill out the information, cut the page out and keep it somewhere you will find it again – your journal, nightstand, Bible, etc.

Keep it as a reminder of the kind of man you are becoming and the lessons you want to make sure you keep with you.

FINAL ENCOURAGEMENT
In Revelation 12:11 (NIV), God talks about those people who overcome the enemy and are able to stand in the final days. It says,

> *"They triumphed over [Satan] by the blood of the Lamb and by the word of their testimony; they did not love their lives so much as to shrink from death."*

Be the kind of man who relies on Jesus every day of your life. The story that God is writing with your life has incredible power and can change the world. And when your life is over, you will know that you did your part well and will be ready for the next adventure God has for you.
You are called to be a common man who lives an uncommon life.

Top 3 Things I Learned and Want to Still Be Applying a Year from Now:

1) _____

2) _____

3) _____

My Coach Is:

_____

I Am Coaching:

_____

My Swim Buddy Is:

_____

My Top Distractions or Temptations Are:

_____

_____

_____

_____

My Top Big Picture Goal Is:

Goal:

Motivation:

Potential Roadblock:

Strategy:

"I want to be the kind of man who_____"

ADDITIONAL NOTES